Adam E. M. Eltorai

Effects of Observation

A Study of Captive Black-Tailed Prairie Dog Behavior

LAP LAMBERT Academic Publishing

Impressum / Imprint

Bibliografische Information der Deutschen Nationalbibliothek: Die Deutsche Nationalbibliothek verzeichnet diese Publikation in der Deutschen Nationalbibliografie; detaillierte bibliografische Daten sind im Internet über http://dnb.d-nb.de abrufbar. Alle in diesem Buch genannten Marken und Produktnamen unterliegen warenzeichen-, marken- oder patentrechtlichem Schutz bzw. sind Warenzeichen oder eingetragene Warenzeichen der jeweiligen Inhaber. Die Wiedergabe von Marken, Produktnamen, Gebrauchsnamen, Handelsnamen, Warenbezeichnungen u.s.w. in diesem Werk berechtigt auch ohne besondere Kennzeichnung nicht zu der Annahme, dass solche Namen im Sinne der Warenzeichen- und Markenschutzgesetzgebung als frei zu betrachten wären und daher von jedermann benutzt werden dürften.

Bibliographic information published by the Deutsche Nationalbibliothek: The Deutsche Nationalbibliothek lists this publication in the Deutsche Nationalbibliografie; detailed bibliographic data are available in the Internet at http://dnb.d-nb.de. Any brand names and product names mentioned in this book are subject to trademark, brand or patent protection and are trademarks or registered trademarks of their respective holders. The use of brand names, product names, common names, trade names, product descriptions etc. even without a particular marking in this works is in no way to be construed to mean that such names may be regarded as unrestricted in respect of trademark and brand protection legislation and could thus be used by anyone.

Coverbild / Cover image: www.ingimage.com

Verlag / Publisher:
LAP LAMBERT Academic Publishing
ist ein Imprint der / is a trademark of
OmniScriptum GmbH & Co. KG
Heinrich-Böcking-Str. 6-8, 66121 Saarbrücken, Deutschland / Germany
Email: info@lap-publishing.com

Herstellung: siehe letzte Seite /
Printed at: see last page
ISBN: 978-3-8473-3881-9

Zugl. / Approved by: St. Louis, Washington University in St. Louis, Honors Thesis, 2010

Effects of observation:
A study of captive black-tailed prairie dog behavior

Adam E. M. Eltorai
Washington University in St. Louis

Table of Contents

Abstract

Black-tailed prairie dogs *(Cynomys ludovicianus)* are diurnal, omnivorous rodents that live in intricate cities. Black-tailed prairie dog social complexity rivals that of some primates, and, in some respects, resembles the behavior of humans. Due to the rich variety of readily-observable, sophisticated behaviors such as coloniality, infanticide, anti-predator behaviors, kin recognition, cooperation, conflict, and reproductive success, the black-tailed prairie dog is a wonderful model species for the study of behavior. Using a captive population of black-tailed prairie dogs, we were able to quantify the effects of observation on key behaviors. Our findings have potentially significant implications for observational research studies and offer further insights into higher-ordered animal behavior.

Key Words: *Cynomys ludovicianus,* black-tailed prairie dogs, captive, social behavior, vigilance, observation

Introduction

Black-tailed prairie dogs *(Cynomys ludovicianus)* are one of five species of prairie dogs—the other four being, white-tailed, Utah, Gunnison's, and Mexican. Prairie dogs belong to the order of rodents and the squirrel family, Sciuridae. They are diurnal and live in socially-intricate cities. The distribution of black-tailed prairie dog ranges from east of Colorado, Montana, and Arizona to west of Nebraska to south of southern Canada to north of Chihuahua, Mexico. Prairie dog colonies, when undisturbed, stretch for kilometers and can contain thousands of individuals (Hoogland, 1995).

Black-tailed prairie dogs are active throughout the year, except during severe weather when the animals become temporarily inactive (Harlow and Menkens, 1986). Foraging from sunrise to sunset, they eat seeds, stems, roots and leaves of plants. Grasses and other herbaceous plants make up a majority of the prairie dog diet (Kelso, 1939). The species of plant eaten varies with the season and the specific colony's location (King, 1955). Black-tailed prairie dogs grow to about 30 centimeters long and weigh about 700 grams (Hoogland, 1995). They are tan, chubby, and have a whitish belly. The tips of their tails are black. They have short ears and large black eyes.

Colonies are made up of territorial, family groups of prairie dogs. These groups are called coteries (King, 1955; Hoogland, 1995). Coteries can vary in size from one to 26 prairie dogs (Hoogland, 1995). The average coterie consists of "one breeding male, two or three adult females, and one or two yearlings of each sex" (Hoogland, 1995: 106). A typical coterie's territory is about one-third of a hectare. This territory usually contains about seventy burrow entrances. If a coterie contains two adult breeding males, they are usually brothers. Additionally, single males have been observed to be the breeding male in two adjacent coteries. Females tend

4

to spend their entire lives in their birth territories; whereas the males are more prone to nomadic tendencies. Males and females begin breeding in their second year. Males are 10-15% larger than females. Males have never been reported to live longer than 5 years; where as some female have lived to 8 (Hoogland, 1995).

The coterie's territory consists of food resources, a burrow system, and one or more burrow openings. Black-tailed prairie dogs spend more than half of their lives within their burrows. Burrows are passed down from generation to generation, as they are energetically costly to construct. Burrows range from simple pipelines to and intricate tunneling system with several entrances, exits, side branches, intersections, food caches, nurseries, and nesting areas. Burrows aid in protection against predation and weather. Burrow mounds, located at burrow entrances, serve to reduce flooding, aid in ventilating the burrow, and function as platforms for scanning for predators (Hoogland, 1995). An excavation of a captive prairie dog population's burrow system showed that the captive group's burrow had no basic differences from that of their wild counterparts (Egoscue and Frank, 1984).

Black-tailed prairie dogs mate underground. As the females are sexually receptive for less than one day per year, the black-tailed prairie dogs breed only once a year. The average litter size is about 3 pups per female. Approximately 50% of the pups survive to the age of one year (Hoogland, 2001). Adults also have a high mortality rate—of those who survive to the age of two, generally only live for two more years (Hoogland, 1995).

Many species prey on prairie dogs. Their predators include: American badgers, bobcats, coyotes, long-tailed weasels, black-footed ferrets, humans, red foxes, grizzly bears, mountain lions, bull snakes, rattlesnakes, golden eagles, northern harriers, peregrine falcons, prairie falcons, Cooper's hawks, red-tailed hawks, buteo hawks, and accipiter hawks (Hoogland,

1995). To improve their chances of survival, prairie dogs have good vision and sight. Prairie dogs have rods and cons, meaning that they have color vision. More specifically, prairie dogs have dichromatic color vision allowing them to see in the blue-green-yellow spectra. Humans have trichromatic color vision, allowing us to see in blue-green-yellow-red spectra (Jacobs and Pulliam, 1973). Black-tailed prairie dogs can hear in essentially the same range as humans (Heffner et al. 1994).

To further improve their chances of survival, prairie dogs "spend about one-third of their time scanning for predators." And when a predator is detected a "prairie dog commonly warns nearby kin with a loud, repetitious antipredator call" (Hoogland, 1995: 2). In prairie dog coloniality, a reduction in predation may be the single most important benefit (Hoogland, 1981). More individuals allow for increased chances of predator detection. Furthermore, with a larger group, an individual can spend less time being vigilant themselves and spend more time eating. With a larger group there is a decreased chance of a particular individual getting eaten (Berger, 1978). Kildaw (1995) showed through experimental manipulation of black-tailed prairie dog group size that the smaller groups foraged relatively more alertly and risk-adversely.

Estimated 100 years ago to have a population over 5 billion, black-tailed prairie dogs are now considered a candidate species for being listed as threatened or endangered. Humans have destroyed a large portion of the prairie dog grassland habitat and introduced the extremely deadly plague. Labeled as pests in competition with livestock for food, black-tailed prairie dogs are poisoned and shot en masse (Slobodchikoff et al., 2009). Humans are unequivocally responsible for driving prairie dogs towards extinction.

"Prairie dogs have a complex social system, rivaling that of some primates and in some respects resembling the behavior of humans. Perhaps that is why prairie dog exhibits in zoos are very popular, with people sometimes spending hours watching the behavior of the animals" (Slobodchikoff et al., 2009: 43). As a social species, black-tailed prairie dogs frequently interact among themselves. Friendly interactions occur between individuals of the same coterie. Examples of such friendly behaviors include: "play, allogrooming, and mouth-to-mouth contact that resemble kisses" (Hoogland, 1995: 2).

The "greet-kiss" (King, 1955), where the prairie dogs "open up their mouths, and press their tongues together for a brief period of time," can occur between any combination of individuals (Slobodchikoff et al., 2009: 56). Steiner proposed the following self-explanatory hypotheses for why this kiss may occur: Food Information Hypothesis (1975), Individual Recognition Hypothesis (1974), and Dominance Maintenance Hypothesis (1975). However "the picture is still not clear about the functions of a greet-kiss. Perhaps the behavior originated for any of the reasons given by the above hypotheses, and has persisted as a mechanism of social reassurance, somewhat like a human kiss" (Slobodchikoff et al., 2009: 57).

Allogrooming, or mutual grooming, occurs when one individual picks the fleas, lice, and ticks from another individual, consequently reducing the number of parasites on the groomed individual (Slobodchikoff et al., 2009). This is another social behavior in which black-tailed prairie dogs take part.

Variations in how much time individual black-tailed prairie dogs spend doing particular behaviors have been studied. Individually-distinct use of time serves to suggest that black-tailed prairie dogs may have unique personalities, which effectively add to the nuanced prairie dog social system (Loughry and Lazari, 1994).

Less amicable behaviors are also observed. February through April, females are defensive of their nursery burrows and hostile interactions frequently occur. However, when the

7

pups emerge from their subterranean burrows in May, friendly interactions are the coterie norm (Hoogland, 1995). The other form of hostile behavior occurs "[w]hen prairie dogs from different coteries meet, they engage in a flagrant territorial dispute that involves staring, tooth chattering, flaring of the tail, bluff charges, and reciprocal anal sniffing. Territorial disputes commonly persist for more than 30 minutes and sometimes include fights and chases as well" (Hoogland, 1995: 2).

From past studies of this North American ground squirrel, interesting findings have been observed regarding coloniality, infanticide, anti-predator behaviors, kin recognition, cooperation, conflict, and reproductive success. Prairie dogs are known to "have a sophisticated communication system that might outstrip monkeys and apes in its complexity, a system on the verge of language," making them a compelling study subject (Slobodchikoff et al., 2009: 1). Readily observable examples prairie dog behaviors include: the "jump-yip display" (individual stretches vertically, throwing their body as they call), rapid scratching to remove fleas, burrow-mound enhancement, and nest-building by collecting leaves, grass and twigs in their mouths (Hoogland, 1995). Due to the rich variety of salient behaviors, the black-tailed prairie dog is a wonderful model species for the study of behavior.

Prairie dogs are even more fascinating study subjects considering the following observations: non-parental prairie dogs make antipredator calls to warn distant kin; lactating females commit infanticide against the offspring of close kin (sister, daughters)—resulting in the major source of pup death; mothers suckle the offspring of female kin (the same pups they tried to kill earlier); and prairie dogs avoid inbreeding with close kin while copulating with cousins (Hoogland, 1995).

Although the black-tailed prairie dog has been extensively studied in the wild, little literature exists regarding *captive* black-tailed prairie dogs (Smith et al., 1973 appears to be the only captive study). "Visitor effect" studies have focused primarily on primates (i.e. Hosey, 2005; Mitchell et al., 1992) while few studies have observed the black-tailed prairie dogs' response to visitors.

Interesting observations were made in the past captive black-tailed prairie dog observational study (Smith et al., 1973). These authors found that summer months for captive black-tailed prairie dogs are characterized by "considerable disorganization but contained the seed of division into groups" (p. 213). Unlike in the wild, "individuals other than the dominant pair in each zoo coterie probably defend boundaries" (p. 214). These differences are most likely to be due to the relatively small amount of space in the enclosure, the high population density, and the inability to emigrate leading to a high density of individuals. Smith et al. also found that "the social behavior of the zoo prairie dogs is broadly comparable to that of their wild counterparts" (p. 214).

In wild population studies on prairie dogs, "strong effects of sex, parental status and environmental context" played a major role in the time spent in vigilant and foraging behaviors (Loughry, 1993: 23). General findings are that parents are more vigilant, individuals are more vigilant in the mornings and when in taller grasses, and that distance from burrow also affect amount of time individuals spend in a vigilant state (Loughry, 1993). Age also profoundly affects how much time is spent being vigilant. As stated by Loughry, "Upon first emergence, pups were extremely wary and spent most of their time vigilant and little time feeding. As pups aged, they increased time spent feeding and decreased time spent vigilant. Male and female pups behaved similarly" (1992: 206).

9

Interesting observations also have been published on vigilance of captive animals. Reduced antipredator behavior has been observed with the more generations a population of animals spends in captivity (McPhee, 2004). Introduction of other species into enclosures alters animals' vigilance behaviors. And group size affects vigilance behaviors (Hardie and Buchanan-Smith, 1997). Black-tailed prairie dogs have been used as a model for determining whether or not predator training has significant effects on survival rates of formerly captive animals (Shier and Owings, 2006). The relationship between observers at zoos and vigilant behavior has not been studied.

In this study, I investigated the relationship between human observer density and the prairie dogs' "daily behaviors," vigilance, social behaviors, and their distribution within their enclosure. Moreover, I looked at how observer density affected the behavior of the different-aged prairie dogs. Vigilance and social behavior are essential components to the prairie dog story. Furthermore, how the individuals distribute themselves within the enclosure may allow for even further insight into captive prairie dog social behavior and their level of vigilance. This has not been the subject of previous research. Ultimately, I hope the findings from this study will help to further our understanding of the broader question: How does observation affect behavior?

Studying black-tailed prairie dogs in captivity offers a unique opportunity to research a popular study species in an uncommonly studied setting. In this study, I did not limit myself to observing the zoo setting's effect on social organization as was done in the Smith et al. (1973). Moreover, as mentioned earlier, prairie dogs natural distribution is centered on the mid-western states of the United States. Thus, the location of this study, the Saint Louis Zoo, much more closely matches black-tailed prairie dogs natural climate, relative to the previous captive study that occurred in the Philadelphia Zoo.

Methods

Subjects

The Saint Louis Zoo is home to approximately 25 black-tailed prairie dogs. The zoo exhibit consists of mostly adult (determined by large size) prairie dogs and seven 1.5 month-old pups. The adults are likely to range from three to five years of age. The sex of the individuals was not able to be determined. Per the zoo keeper, the initial group of individuals was introduced to the exhibit in 1940. The present day zoo population is descendent from the original group.

Data collection

Data were collected using various methods (Lehner, 1996; Washington Park Zoo and Minnesota Zoological Garden, 1983). A digital watch was utilized to properly record the activities to the minute. All 81 hours of observation were detailed by hand using an appropriate data sheet (see Appendix I, II, III, and IV). All observations of the prairie dogs were above-ground and outside the burrows.

An instantaneous scan sample was taken every 5 minutes of the entire group from 9AM to 7PM on three different days to determine the captive population's activity cycle (see Appendix I for data sheet). The number of individuals that were participating in the following behavioral categories at the time of the scan was noted: allogroom, rest, feeding, vigilance, traveling, and other (see Table 1 for complete Ethogram). Using this sampling method, I was able to observe the animals' distribution within the enclosure at various points during the day; the number of individuals physically touching another individual at a given time; the percentage of time the group spends on a particular activity; and the visitor density at a given time.

An instantaneous scan sample was again taken every 5 minutes for *each region* of the enclosure for a one hour period during morning (9AM-11:59AM), afternoon (12PM-2:59PM) and later afternoon (3PM-7PM). The regions are illustrated in Figure 1. Using this sampling method, I was able to observe what activities were occurring in specific regions at different points of the day (see Appendix II for data sheet used).

Focal scan samples of an adult and a pup were taken every 5 minutes from 9AM to 7PM to see if there were any activity and distributional differences with respect to age (see Appendix I for data sheet used). Focal individuals were followed for as long as possible. When the focal individual disappeared into one of the holes, another focal individual, of the same category (adult, pup) was chosen to be followed. The replacement focal individual was chosen by being the most representative of the focal individual category at the time (i.e. resting adult in Region 4).

To further examine the focal individual categories, I used the all occurrence sampling method to focus on specific vigilant and social behaviors (see Appendix III for data sheet used). I monitored the focal individual category (adult, pup) for two hours during each of the three time periods during the day (morning, afternoon, late afternoon). Using this sample method, I observed the duration and frequency of specific behaviors of interest. When an individual went underground, the next focal categorized individual was chosen by being the most representative of the current behavioral trends. The number of socially close individuals to the focal individual was determined by being within a radius of two feet from the focal individual.

Finally, for one hour during the morning, afternoon and late afternoon, I examined, using an instantaneous scan sample every five minutes, how individuals distributed themselves with respect to direct sunlight and shade (see Appendix IV for data sheet used). I examined and attempted to determine whether warm patches (direct sunlight) have any relationship to the prairie dog distribution within the enclosure. Visitor density may not be the only variable

12

influencing prairie dog distribution. Trees hang over all of the enclosure walls, keeping the wall shaded throughout the day.

Table 1—Ethogram

Category of Behavior	Description of Behavior
Allogroom	Grooming of another individual by way of picking (with paws or teeth directly) and eating of insect/parasite/other findings; social value greater than hygiene upkeep.
Resting	Ranges from bending of head down to lying on stomach or back with arms and legs fully stretched; eye rarely completely closed.
Feeding	This behavior consist of physically chewing of food and browsing for food; browsing behavior occurs as individual moves, head down with mouth near or on ground. Slow-medium paced movement.
Vigilance	Characteristic erect, motionless head stares in a particular direction. The frozen body may be laying, sitting or standing.
Traveling	Consists of the animal running in gallop or full sprint to and from various sites.
Other	This can include: pup wrestling, self-grooming, scaling of back wall, burrowing, nest building, and any other behavior that does not fall into one of the aforementioned behavior categories.

Enclosure

A map of the enclosure can be found in Figure 1. The enclosure is approximately 75 x 40 feet. There are relatively uniformly distributed hole-openings throughout the enclosure. The ground is covered in wood chips. Food (generally: kale, Purina rat chow, sweet potatoes, and carrots; periodically: apples, bananas, and bamboo) is dispersed evenly throughout the enclosure each morning. The surrounding walls are textured rock; there is a climbing path on the back wall that allows for the animals to reach elevated vantage points. There are several logs of various shapes and sizes located in Regions 3, 4, and 5 that offer varied climbing, hiding and resting

opportunities.

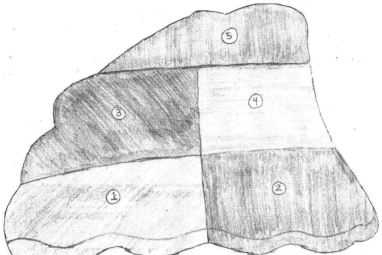

Figure 1. Map of enclosure showing how the space was divided into Regions 1-5.

Statistics

Microsoft Excel was used to generate all of the plots and to determine the lines of best

fit through linear regression. When comparing adults to pups on the same behaviors, only the

regions where overlapping visitor numbers were compared. To determine if adults vs. pups

differed significantly ($p<0.05$) from one another on the same behaviors with respect to increased

visitor density, two-tailed, independent samples Student's t-tests were used for percent of hourly

time spent in each region; percent of hour various social behaviors were exhibited; the average

bout duration of various social behaviors; the average resting, allogrooming, and being vigilant

bout durations; the number of other prairie dogs that are being physically touched; the average

number of other prairie dogs that are socially close; the average number of amiable and agonistic

14

social occurrences; and the percent of time any type of social behavior is displayed. Within subjects ANOVAs were used to see if there are significant differences in regional distribution; in the number of physically touching prairie dogs in each region; in the concentrations of prairie dogs in each of the regions for observations; and in the amount of time adults and pups spend in each of the regions. Two-tailed probability values of the Pearson correlation coefficient r, given the correlation value and the sample size, were found to determine, with respect to increased visitor density, if there were statistically significant changes in the amount of behavior exhibited; the number of physically touching prairie dogs in each region; the duration of behavior bouts exhibited in the number of other prairie dogs individuals are physically touching; the average number of prairie dogs in each region; the percent of hour adults and pups spend in each region; the duration of behavior bouts exhibited; the number of other prairie dogs individuals are socially close to; the on the average number of amiable and agonistic social occurrences; and the percent of time adults and pups display any type of social behavior. Significance levels: $p < 0.05$ (*), $p < 0.01$ (**), $p < 0.001$ (***), not significant (NS). It is critical to note that the behavior observed was significantly more complex than the plotted linear relationship. By plotting the behaviors vs. visitor density in such a straight line fashion, I understand that some of the nuanced changes will be overlooked. However, I believe that plotting the relationships as linear is justifiable for the following two reasons: 1) it simplifies the data to cleanly illustrate the major trends that are occurring; and 2) reduces the differences in the effects of the various types of observers/zoo-goers (i.e. 12 quiet photographers vs. 12 screaming, crying third graders).

Results

Comparison of adult versus pup activity

As a large fraction of the entire population, the adult prairie dogs demonstrate behavior that merits a closer examination. In Figure 2, I show that the adults spend a large portion of their time (41%) resting. Feeding comprises 33% of their overall daily activity. The third largest portion of the adult day is spent being vigilant (15%). The remaining portion of the adult activity is composed of "other" behaviors (6%), traveling (4%), and allogrooming (1%).

If a blanket statement were to be made about pup activity, it would be: Pups are feeding machines. Pups spend 80% of their days feeding, as seen in Figure 2. The pups were never observed performing any "other" behaviors; hence, the "other" category is excluded from the graphs for the pups.

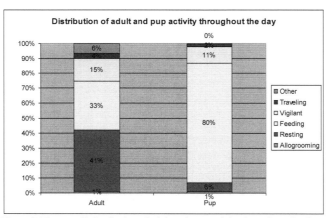

Figure 2. The overall distribution of adult and pup activity throughout the day.

In Figure 3, I describe the adult activity during different times of the day. Feeding becomes increasingly more common as the day progresses; whereas vigilance becomes increasingly less common as the day progresses. Adults spend a significant amount of their days

16

resting—the behavior increases from morning to afternoon and then drops off to just below morning levels in the late afternoon. The other behaviors remain relatively uncommon throughout the day.

To compare, in Figure 3, I describe changing pup activity. Pups spend about four-fifths of their time feeding throughout the day. Vigilance levels are constant during the morning and afternoon but drop off a bit in the late afternoon when feeding behaviors increase slightly. Resting patterns demonstrate that pups have the highest resting levels during the afternoon, followed by the late afternoon resting levels, and finally the morning resting levels are the lowest. Traveling and allogrooming consistently remain rare throughout the day.

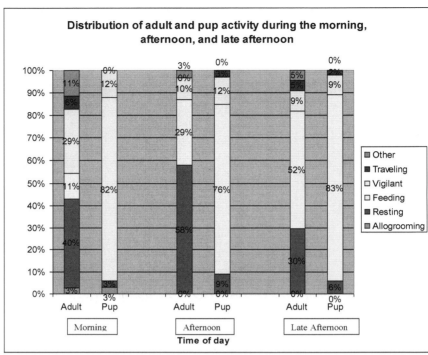

Figure 3. A side-by-side comparison of the adults' and pups' activity during the morning (9AM-11:59AM), afternoon (12PM-2:59PM), and the late afternoon (3PM-7PM).

Comparison of adult versus pup regional distribution

In Figure 4, I show the percentage of time adult prairie dogs spend in each region throughout the day. Adults spend one-third of their day in region 1 and about one-fifth of their day in regions 3 and 4. Sixteen percent of the adult day is spent in region 2 and 8% in region 5.

In comparison, pups spend 28% of their time in region 4, 25% in region 3, 23% in region 1, 17% in region 2, and 7% in region 5 (Figure 4).

18

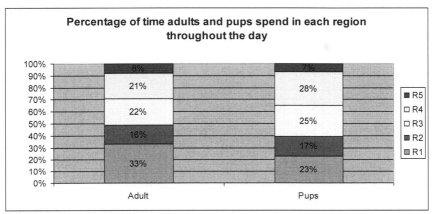

Figure 4. The percentage of time adults and pups spend in each region of the enclosure throughout the day.

In Figure 5, I describe how the percentage of adult time spent in each region changes. The percentage of time adults spend in region 1 decreases throughout the day, whereas the percentage of time adults spend in regions 3 and 4 increases as the day progresses.

In comparison, in Figure 5, I show the percentage of time pups spend in each region as the day progresses. Pups spend an increasingly large percent of their time in region 4. Pups spend a higher percent of their afternoon in regions 2, 3, and 5 than they do their mornings or late afternoons. Pups spend a higher percent of their mornings and late afternoons in region 1 than their afternoon.

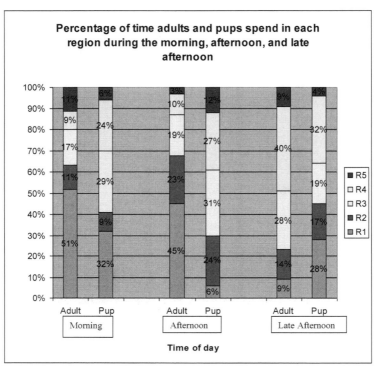

Figure 5. A side-by-side comparison of the percentage of time adults and pups spend in each region of the enclosure during the morning (9AM-11:59AM), afternoon (12PM-2:59PM), and the late afternoon (3PM-7PM).

Differential effects of increasing visitor density on adult and pup activity

In Figures 6, I look at how as the number of visitors increases the adult prairie dog behavior changes. Only one occurrence of adult allogrooming was noted during the full-day focal scan sampling of the adults. The event was recorded during the 10AM hour with 4 visitors present. As the number of visitors increases the percent of each hour adults spend resting increases, feeding decreases, being vigilant decreases, traveling decreases, and performing "other" behaviors decreases.

In comparison, in Figure 6, I look at how as the number of visitors increases the pup prairie dog behavior changes. Only one occurrence of pup allogrooming was noted during the full-day focal scan sampling of the pups. The event was recorded during the 10AM hour with 20 visitors present. No "other behavior" occurrences were noted during the full-day focal scan sampling of pups. As the number of visitors increases the percent of each hour pups spend resting increases, feeding remains constant, being vigilant decreases, and traveling increases. Only 2 occurrences of pup traveling were noted during the full-day focal scan sampling of the pups. The events were recorded during the 1PM hour with 29 visitors present and 4PM hour with 23 visitors.

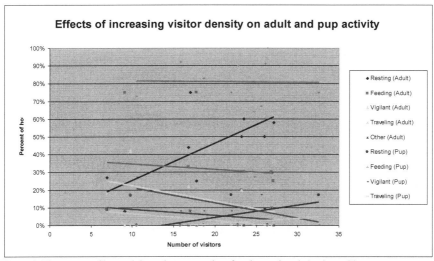

Figure 6. The percent of hour adults and pups spend performing various behaviors with respect to increasing visitor density.

When 10.545 through 27.083 average number of visitors per hour are present, there is a significant difference between percent of hour adults vs. pups spend resting (p=0.000001, t(14)=

8.3681), spend feeding (p= 0.000035, t(14)= 5.9646), but no significant difference in the percent

of hour spent being vigilant or traveling.

Table 2 summarizes the strength and type of the relationships between adult and pup

behaviors and the effect of increasing visitor density.

Behavior	Line of Best Fit Equation	Coefficient of Determination, r^2	Correlation Coefficient, r
Resting (Adult)	y = 0.0209x + 0.0448	0.5059	0.71126
Resting (Pup)	y = 0.0069x – 0.0953	0.3708	0.6089
Feeding (Adult)	y = -0.003x + 0.3753	0.0084	-0.09165
Feeding (Pup)	y = -0.0003x + 0.8178	0.0003	-0.01732
Vigilant (Adult)	y = -0.0087x + 0.3112	0.2029	-0.45044
Vigilant (Pup)	y = -0.0086x + 0.2983	0.2724	-0.5219
Traveling (Adult)	y = -0.0039x + 0.1033	0.2213	-0.4704
Traveling (Pup)	y = 0.003x – 0.0523	0.311	0.5576
Other (Adult)	y = -0.0034x + 0.1223	0.1794	-0.4235

Table 2. Relationship summary between adult and pup behaviors and the effect of increasing visitor density.

In Figure 7, I show that as the number of visitors increases the average duration of

adult allogrooming bouts increases, average duration of adult resting bouts is unchanged, and the

average duration of adult vigilant bouts increases. In Figure 7, I also show that as the number of

visitors increases the average duration of pup allogrooming bouts decreases, average duration of

adult resting bouts increases, and the average duration of adult vigilant bouts decreases.

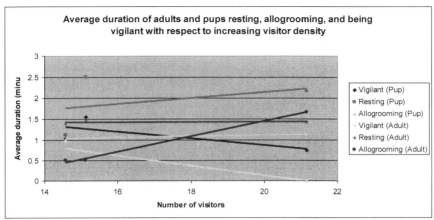

Figure 7. The average duration of adults' and pups' bouts of resting, allogrooming, and being vigilant with respect to increasing visitor density. For clarity, 98 data points are represented by 3 per line.

When 14.58 through 21.16 average number of visitors per hour are present, there is a significant difference in bout duration between adults vs. pups resting (p=0.0001, t(194)= 6.9166), allogrooming (p=0.0001, t(194)= 4.5893), but not for being vigilant.

Table 3 summarizes the strength and type of relationships between behavior duration and the effect of increasing visitor density.

Behavior	Line of Best Fit Equation	Coefficient of Determination, r^2	Correlation Coefficient, r
Resting (Adult)	y = 0.002x + 1.3949	0.0301	0.1734
Resting (Pup)	y = 0.07x + 0.7429	0.1271	0.3565
Allogrooming (Adult)	y = 0.184x -2.2309	0.9943	0.99714
Allogrooming (Pup)	y -0.1234x + 2.5931	0.8122	-0.90122
Vigilant (Adult)	y = 0.0157 + 0.8006	0.2484	0.4983
Vigilant (Pup)	y = -0.0812x + 2.4929	0.5284	-0.7269

Table 3. Relationship summary for the average duration of adults' and pups' bouts of resting, allogrooming, and being vigilant with respect to increasing visitor density.

In Figure 8, I show that as the number of visitors increases the number of other prairie dogs the focal adult physically touches modestly decreases, but the number of other prairie dogs the focal pup touches increases.

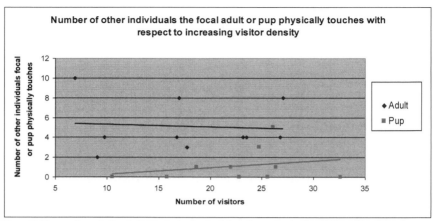

Figure 8. The number of other individuals the focal adult and pup physically touches with respect to increasing visitor density.

When 9.75 through 32.583 average number of visitors per hour are present there is a significant difference between the average number of other individuals adults vs. pups physically touch (p= 0.0004, t(16)= 4.4241).

Table 4 summarizes the strength and type of relationships between number of other physically touching individuals and the effect of increasing visitor density.

Number of others physically touching	Line of Best Fit Equation	Coefficient of Determination, r^2	Correlation Coefficient, r
Adult	$y = -0.0281x + 5.5998$	0.0064	-0.08
Pup	$y = 0.0692x - 0.4574$	0.0665	0.2578

Table 4. Relationship summary between number of other physically touching prairie dogs with respect to increasing visitor density.

Effects of increasing visitor density on activity in individual regions

In Figure 9, I look at prairie dog activity that occurs in each region with respect to increasing visitor density. In region 1, "other" behaviors, traveling, and vigilance rates decrease among prairie dogs as visitor density increases. Feeding in region 1 increases then decreases with

24

the increasing visitor density. Resting decreases initially with increased visitor number but then increases to constitute 55% of the activity in the presence of even more visitors.

In region 2, "other" behaviors increase as visitor number increase. Feeding activity decreases with increasing visitor numbers. Vigilance rates initially decrease then increase dramatically with respect to increasing visitor density. Resting activity is highest at intermediate visitor density levels.

Of note, in region 3, at low visitor density levels, feeding constitutes 71% of the prairie dog activity. As the number of visitors continues to increase feeding behavior decreases then increases again at high visitor numbers. Vigilance remains constant initially then rates increase as the average number of visitors increases. Resting was only observed when the average number of visitors was 13.91.

Feeding levels in region 4 are highest when there is a low density of visitors. Vigilance levels increase as the number of visitors increases. Resting levels are highest when an intermediate number of visitors are present.

In region 5, with respect to increasing visitor density, vigilance levels are highest when there is the smallest number of visitors present. Feeding levels are the highest when there is an intermediate number of visitors. Traveling constitutes a large portion of the observed—up to 25%—particularly with increasing visitor density.

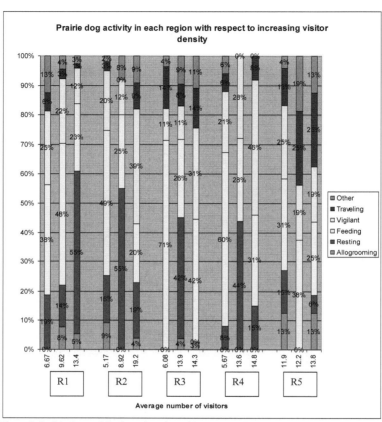

Figure 9. Prairie dog activity in each region with respect to increasing visitor density.

In Figure 10, I look at the number of physically touching prairie dogs in each region as the average number of visitors increases. Visitor number was found by averaging the number of visitors measured every five minutes. The number of physically touching prairie dogs is the sum of the every-five minute regional sampling measurements. With increasing average visitor numbers the number of physically touching prairie dogs in region 1 increases, in region 2 decreases, in region 3 increases, in region 4 decreases, and in region 5 decreases.

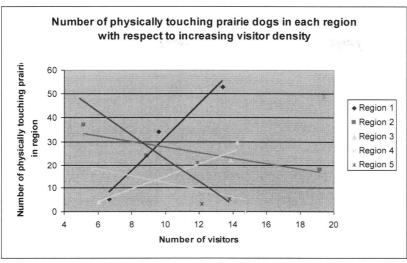

Figure 10. The number of physically touching prairie dogs in each region with respect to increasing visitor density. For clarity, 40 data points are represented by 3 per line.

There is a significant difference between number of physically touching prairie dogs in Region 1 vs. 4 (p= 0.040, F(15)= 5.113), Region 1 vs. 5 (p= 0.039, F(15)= 5.176), Region 2 vs. 4 (p= 0.004, F(15)= 11.967), and Region 2 vs. 5 (p= 0.004, F(15)= 11.594). There are no significant differences between for the number of physically touching prairie dogs in each region for Region 1 vs. Region 2, Region 1 vs. Region 3, Region 2 vs. Region 3, Region 3 vs. Region 4, Region 3 vs. Region 5, and Region 4 vs. Region 5. In sum, there is a significantly more physical touching occurring in the front regions vs. the more distant regions 4 and 5.

Table 5 summarizes the strength and type of relationships between region and number of other physically touching prairie dogs with respect to increasing visitor density.

Region with physical touching	Line of Best Fit Equation	Coefficient of Determination, r^2	Correlation Coefficient, r
Region 1	y = 7.0119x − 38.774	0.9635	0.9815
Region 2	y = -1.1929x + 39.563	0.7935	-0.8907
Region 3	y = 2.7781x − 13.041	0.9297	0.9642
Region 4	y = -1.5257x + 27.322	0.6798	-0.8245
Region 5	y = -4.961x + 72.39	0.2713	-0.5209

Table 5. Relationship summary between the number of physically touching prairie dogs in each region with respect to increasing visitor density.

Effects of increasing visitor density on entire population's regional distribution

In Figure 11, I show that as visitor density increases, the number of prairie dogs in region 1 increases, in region 2 increases, in front regions (regions 1 and 2) increases, in region 3 remains constant, in region 4 decreases, in intermediate regions (regions 3 and 4) decreases, and in region 5 decreases.

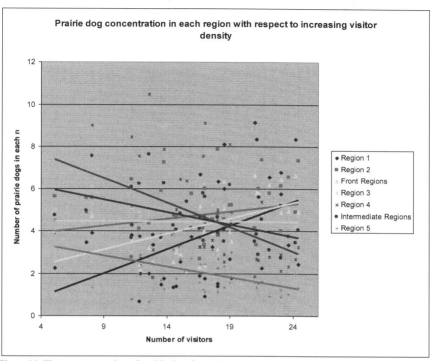

Figure 11. The average number of prairie dogs in each region with respect to increasing number of visitors.

When 5.1 through 24.44 average number of visitors per hour are present there are significant differences between the number of prairie dogs Region 1 vs. 2 (p= 0.023, F(77)= 5.394), Region 1 vs. Intermediate Regions (p= 0.045, F(77)= 4.144), Region 1 vs. 5 (p= 0.000111, F(77)= 16.621), Region 2 vs. 5 (p< 0.0000009, F(77)= 85.789), Region 3 vs. 5 (p< 0.0000009, F(77)= 104.505), Region 4 vs. 5 (p< 0.0000009, F(77)= 44.638), Intermediate Regions vs. Region 5 (p< 0.0000009, F(77)= 105.688). There are no significant differences Region 1 vs. 3, Region 1 vs. 4, Region 2 vs. 3, Region 2 vs. 4, Region 2 vs. Intermediate Regions, and Region 3 vs. 4.

29

Table 6 summarizes the strength and type of the relationships observed.

Region	Line of Best Fit Equation	Coefficient of Determination, r^2	Correlation Coefficient, r
Region 1	y =0.2236x – 0.0174	0.1961	0.4428
Region 2	y = 0.0675x + 36313	0.0431	0.2076
Front Regions	y = 0.1455x + 1.8069	0.2636	0.5134
Region 3	y = -0.0028x + 4.4943	0.0002	-0.0141
Region 4	y = -0.2322x + 8.5783	0.2301	-0.4796
Intermediate Regions	y = -0.1175x + 6.5363	0.2308	-0.4804
Region 5	y = -0.1026x + 3.7633	0.2154	-0.4641

Table 6. Relationship summary of average number of prairie dogs in each region with respect to increasing number of visitors.

Effects of increasing visitor density on adults' and pups' regional distribution

In Figure 12, I show the effects of visitor density on adults and pups regional distributions. When 6.91 through 32.583 average number of visitors per hour are present, there is no significant difference between percent of hour adults vs. pups spend in any of the regions; indicating that throughout the day adults and pups, on average, spend the same number of minutes per hour in each of the regions.

There is a significant difference between percent of hour adults spend in Region 1 vs. Region 5 ($p= 0.016$, $F(17)= 7.266$) and Front Regions as a whole vs. Region 5 ($p= 0.012$, $F(17)= 8.122$). There is a significant difference between percent of hour pups spend in Region 1 vs. Region 5 ($p= 0.042$, $F(17)= 4.903$), Front Regions as a whole vs. Region 5 ($p= 0.012$, $F(17)= 7.970$), Region 3 vs. Region 5 ($p= 0.002$, $F(17)= 12.915$), Region 4 vs. Region 5 ($p= 0.017$, $F(17)= 7.042$), and Intermediate Regions as a whole vs. Region 5 ($p< 0.0009$, $F(17)= 27.479$).

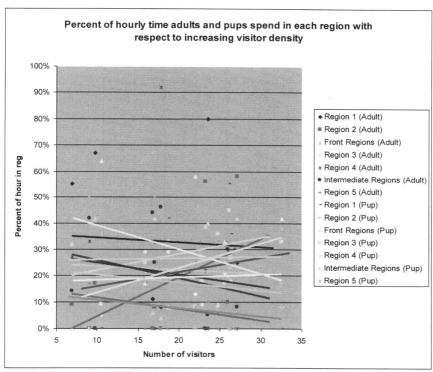

Figure 12. The percent of hour adults and pups spend in each region with respect to increasing number of visitors.

Table 7 summarizes the strength and type of the relationships observed.

Region (Adult/Pup)	Line of Best Fit Equation	Coefficient of Determination, r^2	Correlation Coefficient, r
Region 1 (Adult)	y = -0.0019x + 0.3641	0.0027	-0.0519
Region 1 (Pup)	y = 0.0054x + 0.1041	0.0307	0.1752
Region 2 (Adult)	y = 0.0146x – 0.1011	0.2247	0.4740
Region 2 (Pup)	y = -0.003x + 0.2332	0.0186	-0.1364
Front Regions (Adult)	y = 0.0061x + 0.1582	0.0654	0.2557
Front Regions (Pup)	y = 0.001x + 0.1723	0.0042	0.0648
Region 3 (Adult)	y = -0.0014x + 0.2481	0.0015	-0.0387
Region 3 (Pup)	y = 0.0094x + 0.0384	0.2356	0.4853
Region 4 (Adult)	y = -0.0069x + 0.3259	0.0307	-0.1752

Region 4 (Pup)	y = -0.0093x + 0.4843	0.0709	-0.2662
Intermediate Regions (Adult)	y = -0.0046x + 0.2947	0.0325	-0.1802
Intermediate Regions (Pup)	y = 0.0005x + 0.2547	0.0023	0.0479
Region 5 (Adult)	y = -0.0043x + 0.1601	0.1489	-0.3858
Region 5 (Pup)	y = -0.0031x + 0.1386	0.0430	-0.2074

Table 7. Relationship summary of the percent of hour adults and pups spend in each region with respect to increasing number of visitors.

Role of shade-sunlight on prairie dog distribution

Broadly speaking, the prairie dogs appear to distribute themselves relatively randomly within the front and intermediate regions, I examined if other factors could be dictating the prairie dog regional distribution. I observed how the prairie dogs distributed themselves with respect to sunlight and shade. In Figures 13-14, I look at the relationship of shade and sunlight on prairie dog distribution. All the walls are covered by shade, due to overhanging trees for essentially the entirety of the day.

In Figure 13, I show the changing distribution of prairie dogs throughout the day with respect to their position in the sun and shade (open or against wall). Prairie dogs distribute themselves in decreasing level against the wall as the day progresses. During the afternoon, a majority of the prairie dogs expose themselves to direct sunlight in open areas of the enclosure. In the morning, a majority of the prairie dogs distribute themselves among the shaded regions of the open areas of the enclosure.

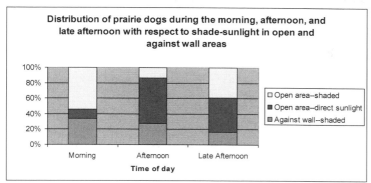

Figure 13. Prairie dog distribution during the morning, afternoon, and late afternoon with respect to shade-sunlight in open and against wall areas.

In Figure 14, I show visible above-ground prairie dog distribution with respect to only sunlight (and consequently also percent of time in shade = 100%-percent of time in sunlight). A majority of the prairie dogs distribute themselves in the shade during the morning and late afternoon, while a majority of the prairie dogs distribute themselves in sunlight during the afternoon.

The next logical question to investigate was what influences such shade-sunlight distribution patterns. To examine this question, I observed sunlight patterns within the enclosure. In Figure 14, I also show the average number of regions within the enclosure—as a percentage of enclosure with direct sunlight—that experience direct sunlight during the morning, afternoon, and late afternoon. Using a *paired* t-test, I observed a highly significant relationship between the distribution of prairie dogs in sunlight and that of the percent of the enclosure with direct sunlight (p= 0.0073, t(2)= 11.6019). Thus I may conclude that the prairie dogs' pattern of distribution within the enclosure follows that of the sunshine. In other words, regardless of their proximity to the visitors, prairie dogs' regional distribution appears to be dictated by sunlight patterns.

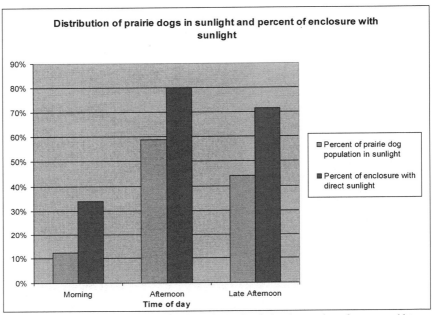

Figure 14. Side-by-side comparison of prairie dog distribution during the morning, afternoon, and late afternoon with respect to sunlight, along with the percentage of enclosure with direct sunlight.

Differential effects of increasing visitor density on adult and pup social behaviors

In Figures 15-19, I look at the relationship between increasing visitor density and adult and pup social behaviors. In Figure 15, I show the percent of adult time performing various social behaviors with respect to increasing visitor density. The adults are alone for approximately half of the time and spend more time being alone with increasing visitor density. As the number of visitors increases the percent of time adults are socially close to others without touching decreases. The percent of time adults are only physically touching without being socially close to others increases as visitor density increases.

Additionally, I show the percent of time pups perform various social behaviors. Pups spend a majority of their time being alone when the number of visitors is fewer. The percent of

34

time pups are socially close but not touching others increases as the number of visitors increases from its lowest level. Pups consistently do not spend much time physically touching others, regardless of the visitor density.

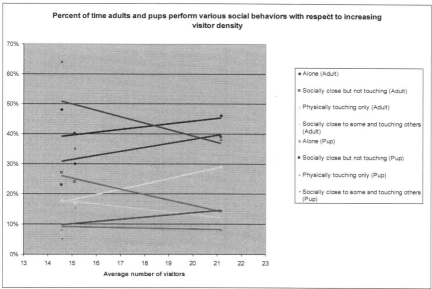

Figure 15. Percent of time adults and pups perform various social behaviors with respect to increasing visitor density. For clarity, 72 data points are represented by 3 per line.

When 14.58 through 21.16 average number of visitors per hour are present there is a significant difference between percent of each hour adults vs. pups spend being socially close to other prairie dogs without physically touching any others (p= 0.0062, t(16)= 3.1488) and the percent of each hour spent touching other prairie dogs (p< 0.0001, t(16)= 5.5936), but no significant difference for the percent of each hour spent being alone or being socially close to some and touching others.

Table 8 summarizes the strength and type of the relationships between adult and pup behaviors and the effect of increasing visitor density.

Social Behavior	Line of Best Fit Equation	Coefficient of Determination, r^2	Correlation Coefficient, r
Alone (Adult)	y = 0.0092x + 0.2576	0.1154	0.3397
Alone (Pup)	y = -0.0211x + 0.8149	0.2339	-0.4836
Socially close but no touching (Adult)	y = -0.0184x + 0.5294	0.9786	-0.9892
Socially close but no touching (Pup)	y = 0.0136x + 0.1097	0.2701	0.5197
Touching only (Adult)	y = 0.018x – 0.0924	0.9775	0.9886
Touching only (pup)	y = -0.0014x + 0.1099	0.1878	-0.4334
Socially close and touching (adult)	y = -0.0088x + 0.3054	0.0746	-0.2731
Socially close and touching (Pup)	y = 0.0073x – 0.0111	0.2367	0.4865

Table 8. Relationship summary between adult and pup social behaviors and the effect of increasing visitor density.

In Figure 16, I show adults' and pups' average duration of performing various social behaviors with respect to increasing visitor density. Duration of bouts of being socially close to some without touching along with bouts of being socially close to some and touching others decrease as the number of visitors increases. The duration of bouts of adults being alone and bouts of only physically touching increase as the number of visitors increases. The pups' average duration increased for all of the examined types of social behaviors with increased visitor density.

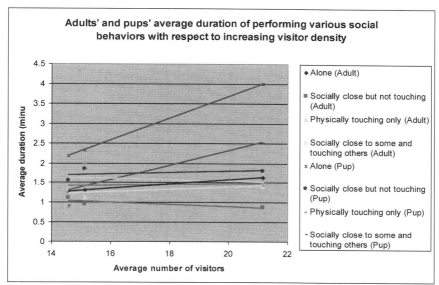

Figure 16. The average duration of adults and pups performing various degrees of social behaviors with respect to increasing visitor density. For clarity, only 3 data points per line are shown.

When 14.58 through 21.16 average number of visitors per hour are present there is a significant difference between the durations of adults vs. pups being alone (p<0.0001, t(88)= 9.3870), being socially close but not touching (p<0.0001, t(36)= 16.8067), but no significant for physically touching only nor for being socially close to some and physically touching others at the same time.

Table 9 summarizes the strength and type of relationships between behavior duration and the effect of increasing visitor density.

Social Behavior	Line of Best Fit Equation	Coefficient of Determination, r^2	Correlation Coefficient, r
Alone (Adult)	$y = 0.056x + 0.4511$	0.9900	0.9949
Alone (Pup)	$y = 0.276x - 1.8404$	0.9990	0.9994
Socially close but no touching (Adult)	$y = -0.0262x + 1.4158$	0.6342	-0.7963
Socially close but	$y = 0.0198x + 1.3999$	0.2107	0.4590

37

no touching (Pup)			
Touching only (Adult)	y = 0.0293x + 0.7883	0.5229	0.7231
Touching only (pup)	y = 0.0112x + 1.2406	0.0052	0.0721
Socially close and touching (adult)	y = -0.0197x + 1.9178	0.9990	-0.9994
Socially close and touching (Pup)	y = 0.1872x -1.4246	0.7214	0.8493

Table 9. Relationship summary between the average duration of adults and pups performing various degrees of social behaviors with respect to increasing visitor density.

In Figure 17, I show that the average number of other prairie dogs that adults and pups are socially close to decreases as the number of visitors increases.

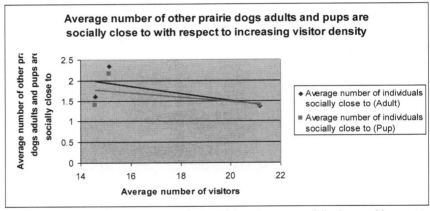

Figure 17. The average number of other prairie dogs adults and pups are socially close to with respect to increasing visitor density. For clarity, 98 data points are represented by 3 per line.

When 14.58 through 21.16 average number of visitors per hour are present, there is no significant difference between average number of other prairie dogs adults vs. pups are socially close to with respect to increasing visitor density.

Table 10 summarizes the strength and type of relationships between number of other socially close individuals and the effect of increasing visitor density.

Number of others socially close	Line of Best Fit Equation	Coefficient of Determination, r^2	Correlation Coefficient, r
Adult	$y = -0.0868x + 3.2387$	0.3955	-0.6288
Pup	$y = -0.0529x + 2.5394$	0.1916	-0.4377

Table 10. Relationship summary between number of other socially close prairie dogs with respect to increasing visitor density.

Kissing and perianal sniffs are used both as greetings and a method for identification. In the wild, if an individual from another coterie is encountered a bout of agonistic behavior is likely to follow. In Figure 18, I show that with increasing visitor density the adults' average number of mouth-to-mouth kisses per hour increases; the average number of perianal sniffs per hour decreases slightly; and the number of adult skirmishes decreases. The data are based on the number of recorded events in which adults were undoubtedly in dispute. The skirmish behavior was classified by any combination of the following: physical attacks, biting, wrestling, and teeth-bearing. In Figure 18, I also show that with increasing visitor density pups' average number of kisses and average number of perianal sniffs per hour decreases, and the number of occurrences of pup wrestling very modestly increases. Two important notes should be addressed: 1) pup wrestling can either be viewed as agonistic or playful and 2) the total number of pup wrestling occurrences is quite a bit higher than the total number of adult skirmishes (43 vs. 12, respectively).

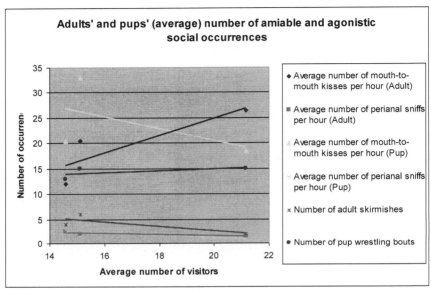

Figure 18. With respect to increasing visitor density, the adults' and pups' average number of mouth-to-mouth kisses and number of perianal sniffs per hour; the number of adult skirmishes; and the number of pup wrestling bouts. For clarity, 48 data points are simplified to 3 per line.

When 14.58 through 21.16 average number of visitors per hour are present there is a significant difference between number of adult skirmishes vs. number of pup wrestling bouts (p<0.0001, t(14)= 12.6557) with adults being "more social." There are no significant differences between average number of mouth-to-mouth kisses per hour or average number of perianal sniffs per hour adult vs. pup with respect to increasing visitor density.

Table 11 summarizes the strength and type of relationships between number of other socially close individuals and the effect of increasing visitor density.

Behavior	Line of Best Fit Equation	Coefficient of Determination, r^2	Correlation Coefficient, r
Mouth-to-mouth kisses (Adult)	$y = 1.7042x - 9.2304$	0.7291	0.8538
Mouth-to-mouth kisses (Pup)	$y = -1.1721x + 43875$	0.2965	-0.5445

40

Perianal sniffs (Adult)	y = -0.1234x + 4.0931	0.8122	-0.9012
Perianal sniffs (Pup)	y = -0.2126x + 5.4384	0.5560	-0.7457
Adult skirmishes	y = -0.4525x + 11.673	0.6821	-0.8259
Pup wrestling bouts	y = 0.1783x + 11.309	0.3179	0.5638

Table 11. Relationship summary of the average number of amiable and agonistic social occurrences per hour with respect to increasing visitor density.

In Figure 19, I show that the percent of time adults display any type of social behavior decreases slightly with respect to increasing visitor density, while the percent of time pups display any type of social behavior increases with increasing visitor density. Any social behavior includes socially close but not touching other prairie dogs, physically touching other prairie dogs, and socially close to some prairie dogs while physically touching others.

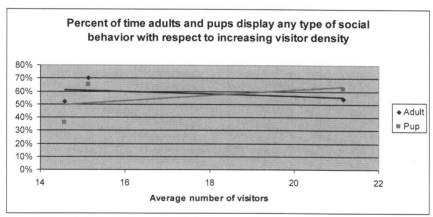

Figure 19. Percent of time adults and pups display any type of social behavior with respect to increasing visitor density. For clarity, 98 data points are represented by 3 per line.

When 14.58 through 21.16 average number of visitors per hour are present there is a significant difference between adults vs. pups for the percent of time any type of social behavior is demonstrated (p= 0.0231, t(94)= 2.2906) with adults being "more social."

41

Table 12 summarizes the strength and type of relationships between number of other socially close individuals and the effect of increasing visitor density.

Displaying any type of social behavior	Line of Best Fit Equation	Coefficient of Determination, r^2	Correlation Coefficient, r
Adult	y = -0.0092x + 0.7424	0.1154	-0.3397
Pup	y = 0.0211x + 0.1851	0.2339	0.4836

Table 12. Relationship summary between percent of time adults vs. pups display any type of social behavior with respect to increasing visitor density.

Discussion/Conclusion

Overall activity

Observations of the Saint Louis Zoo's black-tailed prairie dogs revealed that overall the captive population spends a large proportion of its time feeding, resting, and being vigilant. Upon closer examination, these behaviors change as the day progresses and with respect to the age of the individuals performing the behaviors.

Overall, adults spend the largest portion of their day resting. Feeding is the second most common activity, followed by being vigilant. Pups, on the other hand, spend almost all of their time feeding.

Regional distribution

Next I observed how the population distributed itself within its enclosure. The black-tailed prairie dogs spend 90% of their overall time in front and intermediate regions of the enclosure. As with overall activities, closer examination revealed that prairie dog distribution within the enclosure changes as the day progresses and with respect to the age of the age of the individual prairie dog age.

Broadly, adults spend more than the population average amount of time in closest to the front; whereas, in comparison, the pups spend an increased amount of time towards the back.

Visitor distribution

Once the background data regarding the prairie dogs' behavior and region distribution had been addressed, I began addressing a subject of interest: the effect of observation on behavior. I used the number of visitors standing in front of the prairie dog enclosure as the quantifiable measure of the amount of observation that was occurring. Broadly, the number of visitors to the prairie dog exhibit was, on average, highest during the afternoon. The morning and late afternoon number were less but similar to one another.

Effects of observation on behavior

To explore the relationship of observation and behavior, I began by examining how increases in visitor numbers related to the entire captive population's performance of specific behaviors. Although the absolute amount of time varied, for both adults and pups the amount of time spent resting increased, feeding decreased, and interestingly vigilance decreased with increased visitor numbers.

Upon closer examination, the *duration* of resting bouts for both adults and pups increased with increased visitor density. While allogrooming and being vigilant bout duration increased for adults with increased visitor density, the pups demonstrated the opposite effect—decrease bout duration with increased visitor density. With increased visitor numbers, I showed that adults physically touch fewer individuals; whereas pups tend to touch a greater number. Additionally, I showed that with increasing visitors the number of physically touching prairie dogs increases on the left side of the enclosure (Regions 1 and 3) and decreases on the right side (Regions 2, 4, and 5). I hypothesize that this may have to do with the main burrow entrances being located, in higher concentration, on the left side of the enclosure.

Effects of observation on regional distribution

Next, I observed how increases in visitor numbers related to the prairie dogs' regional distribution. I began by examining the impact on the entire captive population. Interestingly, as

44

the number of visitors increases, the prairie dogs generally flood to the front regions (and exiting the more distant regions), becoming closer in proximity to the visitors. This observation may suggest that prairie dogs demonstrate interspecific sociality. Closer examination showed that adults followed this trend, while pups were slightly more hesitant to rush toward the increasing number of visitors.

To examine the effects of other factors on prairie dog regional distribution, I looked at prairie dog distribution in regards to shade and sunlight. The prairie dogs' pattern of distribution within the enclosure follows that of the sunshine. In other words, regardless of their proximity to the visitors, prairie dogs' regional distribution appears to be dictated by sunlight patterns. This finding points to the need to conduct future studies of prairie dog temperature regulation.

Effects of observation on social behavior

Prairie dogs are undeniably social creatures. To examine prairie dog sociality, I observed adult and pup kissing; physical touching; being in social proximity (with and without touching others) versus being alone; and perianal sniffing. Increases in visitor numbers appears to impact adults and pups differently in completely opposite ways in regards to being alone, touching others and being in socially close proximity. Duration of being alone or touching others increases for both adults and pups when visitor numbers increase. An increase in visitor density relates to increases in pups displaying any social behavior and a decrease for adults. With increased observation, adults kiss more and fight less, whereas pups kiss less and fight more.

Future directions

A comparison of pup to adult social behavior reveals different trends in each age category. I hypothesize that these differences are due to learning. In other words, the pups have not completely learned how to properly behave socially. To draw a broader conclusion from the data, it appears that pups demonstrate social behaviors immediately, but "normal" social behavior takes time and is consequently learned. To help to further investigate this question, data could be collected on "teenagers" to see if their social behavior demonstrates a transition phase between pups and adults.

An additional interesting study would be that of thermal regulation. Are the prairie dogs occupying sunlit regions because of the sun's warmth? If so, what is the ideal prairie dog temperature? Such a study could potentially have interesting prairie dog conservation implications.

Summary

With the aim of studying the effects of observation on behaviors, I chose to observe captive black-tailed prairie dogs. Due to their complex and highly social tendencies, prairie dogs are a wonderful model organism to study behavior. Because prairie dogs are an enjoyable species to watch at the zoo, the masses of visitors offered a quantifiable measurement of observation.

I hope the findings from this study have aided to further our understanding of the broader question: How does observation affect behavior? In many aspects of observed prairie dog behavior there appears to be some relationship with visitor density. Several of these relationships appear to be applicable to the entire population of prairie dogs; while others seem to vary based distance from observers and prairie dog age.

Acknowledgements

Special thanks to my advisor Robert W. Sussman, PhD, for his guidance and mentorship. I also thank John Hoogland, PhD, for his encouragement and suggestions for research and the St. Louis zoo keepers for information regarding the prairie dog exhibit's history.

Appendix

I. Data sheet used for instantaneous scan sample of group and of focal individual categories (adult, pup).

Data sheet key: vstrs = number of visitors at time of scan; R1 (R2, R3, R4, R5) = number of individuals or if focal individual was present in Region 1 of enclosure (see Figure 1); algrm = number of individuals or if focal individual was allogrooming; rst = number of individuals or if focal individual was resting; fd = number of individuals or if focal individual was feeding; vg = number of individuals or if focal individual was vigilant; trvl = number of individuals or if focal individual was traveling; phys tch = number of individuals physically touching or number of individuals the focal individual was touching; othr = number of individuals or if focal individual was performing a behavior "other" then the aforementioned ones.

Time	weather	vstrs	R1	R2	R3	R4	R5	algrm	rst	fd	vg	trvl	phys tch	othr	notes

II. Data sheet used for instantaneous scan sample for each region (1-5).

Data sheet key: wthr = weather at time of scan; in R = number of individuals in the particular region being examined; adults = number of adults in the region; pups = number of pups in the region; algrm = number of individuals allogrooming in the region of interest; rst = number of individuals resting in the region of interest; fd = number of individuals feeding in the region of interest; vg = number of vigilant individuals in the region of interest; trvl = number of individuals traveling in the region of interest; phys tch = number of individuals physically touching another individual in the region of interest; #vstrs = number of visitor at the time of the scan; #othr Rs = number of individuals in other regions at the time of the scan; othr = number of individuals performing a behavior other than the aforementioned ones.

Time	wthr	in R	adults	pups	algrm	rst	fd	vg	trvl	phys tch	#vstrs	#othr Rs	othr	notes

III. Data sheet used for all occurrences of social and vigilant behaviors for focal individual categories (adult, pup).

Data sheet key: individual = the box was marked to indicate the observation of a new individual; start, stop time = beginning and ending of particularly noted behavior; alone = individual was alone in its behavior; socly cls = number of individuals the focal individual was socially close to; tch = number of individuals the focal individual was touching; vg = box was marked if the individual was behaving vigilantly; rst = box was marked if the individual was resting; algrm = box was marked if the individual was allogrooming another individual; othr = box was marked if the individual was performing a behavior other than the aforementioned ones.

49

individual	start, stop time	alone	socly cls	tch	vg	rst	algrm	othr	notes

IV. Data sheet used to examine how individuals distributed themselves with respect to sunlight and shade.

Data sheet key: wthr = weather at time of scan; wall/shade = number of individuals against the wall that were continuously shaded; open sunlight = number of individuals in sunlight away from the walls, positioned towards the center of the enclosure; open shade = number of individuals in shade away from the walls, positioned towards the center of the enclosure; Rs with bright sunlight = which regions posses some amount of direct, bright sunlight.

Time	wthr	wall/shade	open sunlight	open shade	Rs with bright sunlight

References

Berger J (1978). Group size, foraging, and anti-predator ploys: An analysis of bighorn sheep decisions. *Behavioural Ecology and Sociobiology* 4: 91-99.

Egoscue HJ, Frank ES (1984). Burrowing and denning habits of a captive colony of the Utah prairie dog. *Great Basin Naturalist* 44: 495-498.

Hardie SM, Buchanan-Smith HM (1997). Vigilance in single and mixed-species groups of tamarins (*Saguinus labiatus* and *Saguinus fuscicollis*). *International Journal of Primatology* 18: 217-234.

Harlow H J,Menkens, Jr. GE (1986). A comparison of hibernation in the black-tailed prairie dog, white-tailed prairie dog, and Wyoming ground squirrel. *Canadian Journal of Zoology* 64:793–796.

Heffner RS, Heffner HE, Contos C, Kearns D (1994). Hearing in prairie dogs: transition between surface and subterranean rodents. *Hearing Research* 73: 185-189.

Hoogland JL (1981). The evolution of coloniality in white-tailed and black-tailed prairie dogs (Sciuridae: *Cynomys leucurus* and *C. ludovicianus*). *Ecology* 62: 252-272.

Hoogland JL (1995). *The Black-Tailed Prairie Dog: Social Life of a Burrowing   Mammal.* Chicago: University of Chicago Press.

Hoogland JL (2001). Black-tailed, Gunnison's, and Utah prairie dogs reproduce slowly. *Journal of Mammalogy* 82: 917-927.

Hosey GR (2005). How does the zoo environment affect the behavior of captive primates? *Applied Animal Behavior Science* 90: 107-129.

Jacobs GH, Pulliam KA (1973). Vision in the prairie dog: Spectral sensitivity and color vision. *Journal of Comparative and Physiological Psychology* 84: 240-245.

Kelso LH (1939). Food habits of prairie dogs. *U.S.D.A. Circular* 529: 1-15.

Kildaw SB (1995). The effect of group size manipulations on the foraging behavior of black-tailed prairie dogs. *Behavioural Ecology* 6: 353-358.

King JA (1955). Social behavior, social organization, and population dynamics in a black-tailed prairie dog town in the Black Hills of South Dakota. *Contributions from the Laboratory of Vertebrate Biology, University of Michigan* 67: 1-123.

Lehner PN (1996). *Handbook of ethological methods.* 2nd Ed. UK: Cambridge University Press.

Loughry WJ (1992). Ontogeny of time allocation in black-tailed prairie dogs. *Ethology* 90: 206-224.

Loughry WJ (1993). Determinants of time allocation by adult and yearling black-tailed prairie dogs. *Behaviour* 124: 23-43.

Loughry WJ, Lazari A (1994). The ontogeny of individuality in the black-tailed prairie dogs, *Cynomys ludovicians. Canadian Journal of Zoology* 72: 1280-1286.

McPhee ME (2004). Generations in captivity increases behavioral variance: considerations for captive breeding and reintroduction programs. *Biological Conservation* 115: 71-77.

Mitchell G, Tromborg CT, Kaufman J, Bargabus S, Simoni R, Geissler V (1992). More on the 'influence' of zoo visitors on the behaviour of captive primates. *Applied Animal Behaviour Science* 35: 189-198.

Shier DM, Owings DH (2006). Effects of predator training on behavior and post-release survival of captive prairie dogs (*Cynomys ludovicianus*). *Biological Conservation* 132: 126-135.

Slobodchikoff CN, Perla BS, Verdolin JL (2009*). Prairie Dogs: Communication and Community in an Animal Society.* Cambridge: Harvard University Press.

Smith WJ, Smith SL, Oppenheimer EC, de Villa JG, Ulmer FA (1973). Behavior of a Captive Population of Black-Tailed Prairie Dogs. Annual Cycle of Social Behavior. *Behaviour* 46: 189-220.

Steiner AL (1974). Body-rubbing, marking, and other scent-related behavior in some ground squirrels (Sciuridae): A descriptive study. *Canadian Journal of Zoology* 52: 889-906.

Steiner AL (1975). "Greeting" behavior in some sciuridae, from ontogenetic, evolutionary, and socio-behavioral perspective. *Naturalist Canadian* 102: 737-751.

Washington Park Zoo, Minnesota Zoological Garden (1983)."Research Methods for Studying Animal Behavior in a Zoo Setting." Portland.

Printed in Great Britain
by Amazon

84214578R00036